Artemis I Mission

What You Need To Know About NASA Artemis I Launch

Dominic L. Bell

Copyright – All rights reserved. No part of this publication may be reproduced, distributed, or transmitted in any forms or by any means, including photocopying, recording or other electronic or mechanical methods, without the prior written permission of the publisher, except in the case of brief quotations embodied in critical reviews and certain other non commercial users permitted by copyright law.

TABLE OF CONTENTS

Chapter 1 - What is Artemis I

Chapter 2 - Where is Artemis 1 going?

Chapter 3 - WHAT is special about the mission rocket called the Space Launch System (SLS)

Chapter 4 - How will Artemis 1 contribute to science

Chapter 5 - Why has Artemis 1 been so delayed?

Chapter 6 - What are the next steps?

Chapter 1 - What is Artemis I

The first unmanned test flight of NASA's Artemis program, formally known as Artemis I and once known as Exploration Mission-1, is Artemis 1. Additionally, the entire Orion spacecraft and the agency's Space Launch System rocket both make their debut during this launch.

The test mission will take off from Launch Complex 39B at the Kennedy Space Center on August 29, 2022, the same day that Apollo 10 was launched 53 years earlier.

All the rocket stages and spacecraft that will be utilized in later Artemis missions will be tested during the three-week-long Artemis 1 mission.

The mission will discharge ten CubeSat satellites after reaching orbit and executing a trans-lunar injection (burn to the Moon), and the Orion spacecraft will then enter a far-off retrograde orbit for six days. Returning to Earth, the Orion spacecraft will reenter the atmosphere while being shielded by its heat shield and splash down in the Pacific Ocean.

All kinds of achievements for human spaceflight are about to be broken thanks to NASA's new moon program. This project, which is named after Apollo's twin sister, the Greek goddess Artemis, will send a woman and a person of color to the moon for the first time.

If everything goes according to plan, these astronauts will be the first people since Gene Cernan and Harrison Schmitt of Apollo 17 to walk on the lunar regolith (also known as the dusty moon dirt) in 2025.

The Artemis program will also build the first permanent human settlement on the moon by launching a space station into orbit and setting up a base camp there. By taking these steps, we will be able to send astronauts to Mars in the long term, which will be yet another first.

However, before all of that can happen, the space agency needs to test its tools with a mission called Artemis 1 that will set new milestones. What you need to know about the program making big news is provided here as NASA's enormous Space Launch System (SLS) rocket prepares to launch on this historic mission.

Chapter 2 - Where is Artemis 1 going?

Orion, a spaceship that will orbit the moon and eventually bring human crew members there, will be put through its tests during the 42-day Artemis 1 mission. No earlier than August 29 at 8:33 a.m. Eastern time, the unmanned mission will launch from Cape Canaveral, Florida, with September 2 and September 5 serving as backup dates.

Orion will enter the atmosphere in Earth's orbit and then launch into space using the Interim Cryogenic Propulsion Stage (ICPS), a 45-foot-long, cylinder-shaped vehicle with a single engine. A servicing module provided by the European

Space Agency will make any necessary course corrections as Orion glides toward the moon.

In lunar orbit, the spacecraft will make up to one and a half rotations, breaking the previous record for the most distance covered by a crew-carrying spacecraft.

The moon's gravity will then help accelerate it back toward Earth at precisely the appropriate moment after it has fired its engines.

The fastest reentry of any human spacecraft will be made on October 10 when the Orion spacecraft roars back into our atmosphere at a speed of 6.8 miles per second. It is essential to this test flight that the craft and its heat shield withstand temperatures of 5,000 degrees Fahrenheit because NASA cannot recreate these conditions on Earth.

Orion will touch down in the Pacific Ocean near San Diego, California, if it survives, and will be visible to a U.S. Navy ship that will retrieve it.

Although NASA does not yet have a firm plan for a crewed mission to Mars as of 2022, the Artemis 1 flight is usually advertised as the start of Artemis's "Moon to Mars" program.

In order to increase public awareness, NASA created a website where anybody can get the mission's digital boarding pass. The names you submit will be stored on a hard drive that will be positioned inside the Orion spacecraft. A digital copy of the 14,000 essays submitted for the Moon Pod Essay Contest, which was organized by Future Engineers for NASA, will also be on board the spacecraft.

Chapter 3 - WHAT is special about the mission rocket called the Space Launch System (SLS)

The Space Launch System (SLS), developed by NASA, is a super heavy-lift launch vehicle that lays the groundwork for human exploration outside of Earth's orbit. The only rocket that can deliver Orion, humans, and cargo directly to the Moon on a single mission is SLS, which has previously unusual capability and capacity.

SLS, the most powerful rocket in the world, has the capacity to transport more payload into deep space than any other spacecraft since it has a larger payload mass, volume, and energy.

The SLS rocket is made to be adaptable, making it possible to launch more diverse missions, such as robotic scientific expeditions to locations including the Moon, Mars, Saturn, and Jupiter as well as human trips to the Moon and Mars.

The Artemis I mission, NASA's first exploration-class rocket created for manned space flight since the Saturn V, has been delivered and is currently being prepared for by the SLS team. Progress is being made by engineers and business partners in the delivery of rockets for the upcoming Artemis missions.

SLS will develop into ever-stronger variants to meet America's future requirements for deep space missions. Orion or other cargo will be sent to the Moon by SLS, which is nearly 1,000 times farther than where NASA's International Space Station is currently located in low-Earth orbit. SLS is designed for deep space missions. In order to take Orion to the Moon, it must travel at a speed of 24,500 miles per hour, which will be made possible by the powerful rocket.

The Block 1 design, which is used by Artemis I, the first integrated SLS and Orion mission, stands 322 feet tall and weighs 5.75 million pounds. SLS will generate 8.8 million lbs. of maximum thrust during launch and ascent, which is 15% more thrust than the Saturn V rocket.

For Artemis I, Block 1 will send an unmanned Orion spacecraft 280,000 miles from Earth in an orbit 40,000 miles beyond the Moon. Before a manned flight, this mission will show off how well SLS, Orion, and Exploration Ground Systems operate as integrated systems. Astronauts will be flown to the Moon's orbit as part of the Artemis II mission. The Moon-landing mission is made possible by these missions.

The SLS is the most formidable rocket that has ever been created. It is 32 floors high and around 6 million pounds in weight. NASA hired a number of businesses to manufacture it, including Northrop Grumman for the boosters, Aerojet Rocketdyne for the engines, and Boeing for the rocket's orange core stage. The project's estimated cost of $23.8 billion received considerable flak for exceeding the budget.

When the SLS lifts off, it will have a thrust of around 8.8 million pounds, which is significantly more than the 7.5 million pounds of thrust of the Saturn V rocket that launched the Apollo missions.

But when SpaceX's Starship, which is presently under development and is intended to transport people to deep-space destinations, lifts off, it will become the most potent rocket thanks to its staggering 17 million pounds of thrust. NASA notes that the SLS rocket is the only one capable of carrying

Orion, people, and cargo to the Moon in a single mission.

Chapter 4 - How will Artemis 1 contribute to science

Three mannequins will venture into outer space on Artemis 1, even though no humans will fly on it. Their goal is to determine whether future astronauts can safely occupy the Orion spacecraft.

Commander Moonikin Campos, a test subject wearing the Orion Crew Survival System spacesuit, will be in the front of the capsule. The acceleration, vibration, and radiation that the Moonikin is subjected to will be measured by sensors, providing NASA with information about how its human crew members may suffer.

The effects of space radiation on a woman's body will be evaluated using the other two mannequins, Zohar and Helga. The dummies are constructed from plastic slices that represent soft tissue, bones, and lungs. Each will have 5,600 sensors that will keep track of the effects of radiation on the lungs, stomach, uterus, and bone marrow. Helga won't be wearing a protective vest, but Zohar will.

This research is essential as NASA gets ready to send the first woman to the moon. Ramona Gaza, science team leader at NASA's Johnson Space Center, stated in a news briefing that "women, in general, have a higher risk of acquiring cancer as they have more radiation-sensitive organs such as breast tissue and ovaries."

Ten CubeSats, or tiny satellites about the size of a shoebox, will also be carried by Artemis 1. After launching Orion into orbit, the ICPS will separate from the spacecraft and place these satellites at three different positions between Earth and the moon. One of these CubeSats will go to a nearby asteroid using a solar sail so that it can take pictures of it.

A different one has yeast in it to test how space radiation affects living things. The other CubeSats will do additional research, test airbags in a lunar crash landing, image the moon and the spacecraft, scan the moon using a spectrometer, and examine lunar ice.

Chapter 5 - Why has Artemis 1 been so delayed?

The launch of Artemis 1 was initially scheduled for 2016. But a number of things made this objective difficult and time-consuming.

This deadline was impossible due to the Covid-19 outbreak, manufacturing delays for both SLS and Orion, and difficulties getting enough financing from Congress.

In the run-up to the SLS rocket's launch on Monday, NASA had trouble with the wet dress rehearsals or practice runs.

The rocket had three wet dress rehearsal failures in April. The failure of a vent valve and a hydrogen leak were just two of the problems that kept NASA from finishing each test.

In June, NASA's fourth attempt ultimately succeeded after loading the rocket's fuel tanks and running through the countdown from T-29 seconds to the 10 minutes prior to liftoff. NASA considered the test successful despite a second hydrogen leak that caused some of the practice to be delayed.

Chapter 6 - What are the next steps?

Following the completion of the Artemis 1 mission, astronauts will return to the moon for the Artemis 2 and Artemis 3 missions.

Artemis 2 will take a human crew on a lunar flyby after this initial test mission, entering the moon's orbit and returning in eight to ten days. The mission will currently launch in 2024.

If all goes as planned, Artemis 3 might happen as soon as 2025. For the first time in more than 50 years, a crew of astronauts will go to the moon's surface during this mission.

NASA revealed 13 potential areas for Artemis 3 humans to land on the moon last week. All are located close to the lunar south pole, where scientists are focusing their studies. Scientists think frozen water may be present below the surface in the arctic region's perpetually dark and chilly atmosphere.

Depending on the launch date, one of these locations will be the final stop.

The NASA "Moon to Mars" plan, which aims to make the moon a stopover for humans on longer space voyages, is just getting started with the Artemis program.

The Lunar Gateway, an outpost orbiting the moon that will be built in space and support further exploration, will be established by Artemis. The establishment of a lunar base camp by NASA will allow astronauts to reside there while on extended missions and test exploration techniques that might be used on Mars.

Within 20 years, astronauts could be setting foot on Mars by building on Artemis' accomplishments.

www.ingramcontent.com/pod-product-compliance
Lightning Source LLC
Chambersburg PA
CBHW050329220526
45465CB00005B/2193